Above: *Washday in Longor, north Staffordshire. Three generations of the Knowles family in 1900.*
Cover: *The laundry at Shugborough, Staffordshire Arts and Museum Service.*

LAUNDRY BYGONES
Pamela Sambrook

A Shire book

CONTENTS

The washhouse 3
Soaking and soaping 5
After the wash.................................... 9
Washing machines 10
Mangles and wringers........................ 13
Ironing 16
Specialised irons 22
Standing irons................................... 24
Accessories and fittings 29
Further reading 30
Places to visit................................... 32

Published in 2004 by Shire Publications Ltd, Cromwell House, Church Street, Princes Risborough, Buckinghamshire HP27 9AA, UK.
Copyright © 1983 by Pamela Sambrook. First published 1983; reprinted 1987, 1993, 1997 and 2004. Shire Album 107. ISBN 0 85263 648 2.

Printed in Great Britain by CIT Printing Services Ltd, Press Buildings, Merlins Bridge, Haverfordwest, Pembrokeshire SA61 1XF.

ACKNOWLEDGEMENTS

The author wishes to thank the following for permission to reproduce illustrations: Castle Museum, York, pages 2, 6, 21, 30; Hertfordshire Constabulary, page 14; Mary Evans Picture Library, page 5; National Museum of Wales (Museum of Welsh Life), pages 4, 8 (both), 16, 18 (all), 25, 26 (both), 27 (both); Science Museum, South Kensington (Crown copyright), pages 9, 11 (top), 13, 17 (lower), 20 (top), 22, 23 (lower), 24; Staffordshire Arts and Museum Service, pages 1, 3, 7, 11 (lower), 12, 15 (both), 17 (top), 19, 20 (lower), 23 (top), 26 (top), 28, 39, 31 and front cover. Photographic work by Cliff Guttridge.

An extract from the 1929 catalogue of Wilsons & Mathiesons Ltd, Armley, Leeds, showing the Star Corner Copper. The copper's advantages were listed in the accompanying text: it was self-contained, stood in a corner and took up little floor space; it had a good large fire, with three heavy brick linings; it had a galvanised lid with copper hinges and was equipped with a steam escape.

Women prisoners working in the laundry of Stafford Gaol, 1869.

THE WASHHOUSE

Before the nineteenth century washing techniques were simple and traditional. Repeated washing, particularly with soap, was considered highly injurious to clothing, for soap often had a high soda content, dyes might be fugitive, richly coloured silks were expensive to replace if ruined, poor-quality silks were full of body-giving agents which washed out, and the more easily washed cottons were unfashionable except during the Regency period. On the other hand dirt, particularly sweat, quickly rotted clothes that were left unwashed, as can be seen in many silk gowns which have survived intact and glorious but for ragged holes under the arms.

Traditionally, washing was done in cold water, originally at the side of a running stream, later in a wooden tub. Hot water, when used, was heated in an iron pan over an open wood fire. By the late eighteenth century at least three different types of washhouse had developed from such simple beginnings.

Cottages and farms were fitted with enclosed *furnaces* consisting of a copper or cast-iron cauldron built over its own fire, usually in a small outhouse or scullery and often backing on to the main kitchen chimney; in smaller cottages the boiler might be situated in the open backyard over a brick hearth. These were the forerunners of the portable galvanised 'coppers', which sold for between 22 and 36 shillings in the 1890s. Considerably cheaper were the smaller boilers that sat on top of the kitchen range and had a loose base plate to prevent burning. Many a country housewife had nature to help her and little else; her back garden was her washhouse, her hedge a clothes line. Improved facilities depended on a piped water supply; far from such a supply, most country people carried their water from a natural or man-made source

3

which might be up to a mile away, steadying their bucket loads with a home-made wooden yoke lying across the shoulders or over the tops of the buckets.

More elaborate were the laundries of the great country houses. These consisted of a separate washhouse with ironing and mangling rooms. The laundry at Shugborough, Staffordshire, included two main boilers (one for heating the furnace and one for water to be piped around the room), coopered sinks with brass taps, larger coopered tubs for the larger wash, as well as for rinsing and bluing, and a tiled floor sloping to a central drain. The three or four laundrymaids were employed full-time, starting work at four or five o'clock in the morning and working through the household wash in one week. In a smaller house the washerwoman was brought in only periodically and perhaps took home the mangling to do on her own box mangle, which served the village community. She charged perhaps 1d per dozen small items and 2d per dozen larger. By the mid nineteenth century some middle-class families relied on her calling for their dirty clothes and returning them clean. Some modern commercial laundries owe their origins to this type of service, others to specialist laundries which grew up as part of an individual institution such as a convent or workhouse.

Working-class town and village families relied on one of the many communal washhouses built in the nineteenth century, along with a privy and bakehouse, to serve the needs of between twelve and twenty families. These were usually fitted with a sink and pump as well as a furnace. After the 1870s it became more common to incorporate individual washhouses and privies at the rear of the 'tunnel-back' houses, and the communal washhouse declined in importance.

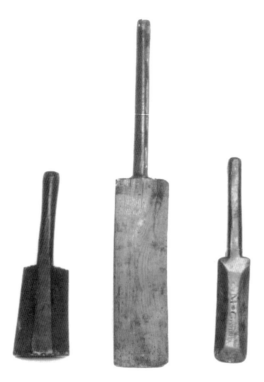

Three washing bats from the collections of the Museum of Welsh Life, St Fagans. These were early, simple washing aids used to pound the clothes and loosen dirt.

4

A woman using a rectangular wooden washing tray.

SOAKING AND SOAPING

Bucking with lye was an old method of washing which was retained in the nineteenth century for certain types of laundry but which previously was universal. It was supplemented and later replaced by hot water washing using soap and some form of hand or mechanical agitator.

Lye was an alkaline liquid obtained from wood ash. The ash, as white as possible, was collected from the furnace, bread oven or open fires and placed in a conical wooden sieve over a tub; water was poured over it and stirred, thus washing out some of the alkaline salts. The resulting water that collected in the tub was filtered through muslin. Ashes varied: oak ash produced the strongest solution, apple wood the whitest wash. Other substances used in lye included pigeon or hen dung, bran and urine.

Lye was used as a preliminary to soaping or on its own, especially for children's clothes and white and table linen. Instead of being rubbed or pounded in a tub, the clothes were folded loosely into a wooden tub called a *buck* (a small tub was called a *bucket*). Sticks were placed crosswise amongst the clothes to prevent them becoming too closely packed. Lye was then slowly poured over the buck and later drawn off at the bottom with a spigot; this was repeated until the lye came through clean. The clothes were then taken out of the buck and rinsed. Badly stained clothes could be boiled in lye.

Lye was also used in soap-making, itself an old-established craft. Fat and some form of alkali were boiled together and then the mixture was precipitated into a curd by adding common salt. Any sort of fat could be used: waste kitchen grease in home-made soap, palm or whale oil in the most expensive commercial product. Alkali was obtained from

5

An extract from the Ironmongers' Standard Catalogue, 1936–7, showing dolly tubs and a wringer stand.

A dolly peg (right), washing punch and long-handled posser. Using a dolly peg is a strenuous task, requiring both a rotational and a vertical movement. The washing punch was used to similar effect but with less of a rotational action. Possers were used with a vigorous vertical movement and usually have a long handle with no crosspiece; small sink versions, sometimes with a crosspiece, were also available. Early possers were home-made of solid or perforated pine; mass-produced ones were made in a variety of patterns usually based on a perforated copper cone that worked at least partly by suction. A galvanised version called the Pearl Wash Pump incorporated a small bellows attachment.

lye or burnt seaweed. Most home-produced soap was soft, kept in bowls or in small round balls. During the nineteenth century soap became cheaper and more available but was often of poor quality and full of soda, hence the housewife's puckered and bleached hands. This also had a deleterious effect upon colours, especially since dyes were fugitive; common salt was used to fix colours.

For stubborn dirt or stains a number of substances were used. A hot flour and water paste was used on general dirt, ground pipeclay on grease spots. Mildew was treated with a mixture of soft soap, ashes and salt; the mixture was rubbed into the stains and the linen laid out on the grass to bleach naturally. The manuscript household recipe book compiled by Mrs Parker-Jervis of Meaford, Stone, Staffordshire, around 1750, contains instructions for the removal of iron mould: 'To the juse of half a lemon, a spoonful of water add 3 drops of camphorated spirits of wine mix'd with oyle of turpentine... Dip the stain in it and lay it wet on a pewter dish or plate and clap a hot iron on it which will fetch it out, then lay it in cold water.'

Sour milk was a well-tried and efficient remover of many stains, especially ink and blood.

Bleaching was usually done by spreading the clothes out on the lawn in a sunny sheltered spot. A night's frost was reputed to improve whiteness.

Even when soap became commonplace, the wash was usually prepared by soaking overnight or longer in lye or cold water with a handful of soda thrown in. The clothes were then rubbed through and wrung before washing in water as hot as possible.

Either a round stave-built tub or a rectangular pine washing tray was used for hand rubbing, often with the aid of a corrugated washboard made of wood, brass, copper, zinc or even glass. Some early washing machines incorporated a corrugated board inside the tub.

Larger quantities of washing using a *dolly peg*, *posser* or *punch* were done in a stave-built *dolly tub*, often bought second-hand from a dyeworks or brewery. Zinc dolly and hand tubs, lighter to carry and easy to clean, became available towards the end of the nineteenth century.

After rubbing, pounding and rinsing, white clothes were usually boiled in a

7

furnace with shredded soap for at least one hour. The clothes were lifted out with a straight *dolly stick* and rinsed, usually in three waters, warm, cool and, third, blued water.

Above: *Washboards were designed for rubbing by hand or scrubbing brush but at least one American invention – the American Washer – was developed to ease the wear and tear on the laundress's hands. The directions for use are explicit: 'Place a plain or ribbed American washboard in the wash tub in as slanting a position as possible. All the soapsuds cover the lowest part of the washboard. With the left hand draw the immersed linen flat as the board. With the right hand roll the machine with slight pressure up and down until the article is cleaned. The washing is quickened by allowing the machine with every movement to take up the largest quantity of suds.'*
Below: *Two glass calenders for polishing starched linen and cotton cuffs, collars or caps. Now in the collections of the Museum of Welsh Life.*

The 'Faithfull' washer, produced in 1906, worked on a hand-rocking principle which produced a figure-of-eight movement of the wash. Lent to the Science Museum, London, by E. Clissett Esq.

AFTER THE WASH

Size, made by boiling hoof clippings, was probably the first stiffening agent used, but the starch used to stiffen high Elizabethan collars and ruffs was obtained by boiling grain; later the best starch was still made from rice but the most common source was old potatoes, which were coarsely grated under water and left to produce a thick paste. The high collars of Victorian men's shirts were steeped in cold-water starch, which included borax to produce a gloss and sometimes turpentine to make the iron glide over the fabric. Articles requiring less stiffening were treated with boiling-water starch, in which bleached beeswax replaced the turpentine. According to Mrs Parker-Jervis, wax was also used on its own: 'Wash in cool suds, put salt in to keep the colours . . . . put white wax into the water before you set it on the fire; half a cake of white wax which costs 7d is enough for one gown; let the gown be dry before you iron it; then iron it on the right side and rub it with a slick stone; the bottom of a bottle will do very well.'

A number of other household substances were used to give body to clothes after washing. They included potato starch sponged on to silk, old tea leaves boiled with nettle tops for plain linens, size from boiled hoof parings for woollens, bran water for chintz, gin or whisky for black or blue silks, and boiled ivy leaves for prints.

Washing was often dried outside on a hedge or flat on the grass or fixed to a line with cleft wooden pegs. Indoors, washing was hung on a slatted wooden rack or even a single length of wood fitted with a pulley to hoist it to the ceiling; alternatively damp washing could be dried and pressed by placing it flat between sheets of paper underneath the hearth rug. Large estate or commercial laundries had sliding drying racks built near to the boiler. Items such as socks and stockings were dried on wooden, china or wire shapes.

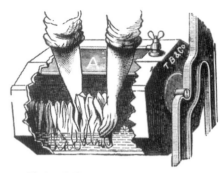

Illustrations of Thomas Bradford's 'Victress Vowel' machine from a handbook of laundry techniques, 'The Art and Practice of Laundry Work' by Margaret Cuthbert Rankin, published by Blackie & Son Ltd in the 1920s.

"Vowel" Washing-Machine

WASHING MACHINES

The first known record of a washing machine appears in the diary of Robert Hooke in 1677, and an early patent was taken out in 1780 by Rodger Rodgerson, but it was left to Victorian inventiveness to produce the numerous hand-driven contrivances, all working towards the desired aim of agitating wash and water without damaging the cloth.

Most early models were of the cradle type, working on a rocking box principle. This was later incorporated into the 'Faithfull' washer, dating from 1906, and into a very simple, open galvanised washer that was still available in the 1940s.

More sophisticated machines appeared in the 1850s and 1860s, notably that of J. Picken of Birmingham, which worked on the same principle as a butter churn. Thomas Bradford of Salford marketed a similar type of machine, the 'Victress Vowel' series, which became the most popular washing machine throughout most of the late nineteenth and early twentieth centuries and was highly recommended by laundry manuals even in the 1920s.

Some of the most successful machines attempted to reproduce mechanically the circular and vertical motion of a hand-operated dolly peg. Perhaps the most successful in reproducing the combined motion of the dolly was a machine made by William Sellers of Keighley around 1890. This incorporated a half-cogged wheel that reversed the dolly and a lug that raised and lowered it.

Yet another type of motion depended on a lever-operated drum fitted inside a square tub. This became the principle on which some of the earliest electrified machines operated.

A number of different types of horizontally mounted agitator were tried: a lever-operated machine made by the CWS, Keighley, incorporated paddles mounted on a horizontal post. A machine now in the Abbey House Museum, Kirkstall, Leeds, and made by E. & F. Turnbull Ltd, Newcastle, used three shaped wooden rollers mounted horizontally.

It is uncertain who produced the first electrified washing machine; claims were made by both the Thor Washing Machine Company (1906) and A. J. Fisher (1908).

Fisher's machine was an electrified dolly-peg type and was extremely dangerous. It must have been similar to a machine in the collections of the York Castle Museum, consisting of an ordinary galvanised dolly tub and a dolly peg rigged up to a belt driven by an electric motor mounted below and to the side of the tub.

A dolly-peg type petrol-driven machine was made by the Maytag Company of Iowa in 1914. Maytag made further developments on the reversing agitator in 1922; by the 1920s electric machines had become more common, but the electric motor still appeared as an afterthought mounted under the tub. Such machines were extremely expensive, costing between £30 and £50. British manufacturers lagged behind their German, American and Canadian colleagues; this was partly because of the cost but also because of the lack of a co-ordinated national electricity grid, which was not set up until the late 1920s. At that time the British market expanded but rapidly became

Right: The 'Red Star' washing machine, made by the Canadian company of Beatty in 1914, worked on the mechanised dolly-peg principle. This was a very successful and popular machine for many years, although it reproduced only the reversing rotational movement of the dolly and not the vertical movement. Lent to the Science Museum, London, by M. Colston Esq.

Below: The 'Gem' washer consisted of a ridged roller used to force dirt out in the same way as a washboard; it was designed to fit on to the side of a sink.

11

dominated by the transatlantic firms of Hotpoint and Thor. In 1927 Thor produced an electric machine with ironing, sausage-making and other attachments. During the early 1920s the first cabinet-type machine and the first tumbler dryer were introduced, both by American companies.

Hand-driven machines continued to be marketed well into the 1940s, while what was probably the first attempt at heating water inside the machine was made by Morton as early as 1880. Before 1920 most machines were made mainly of wood. In the early 1920s, and a little before in the case of some of the Thor machines, a combination of wood and zinc or copper was used. By the later 1920s most machines were metal.

A drum-type copper and wood washing machine with an electric motor mounted underneath. This was supplied by the Thor Harley Machine Company, of Chicago, New York, Toronto and London, before 1920 to the house laundry at Freeford Manor near Lichfield, Staffordshire.

12

The 'Housewife's Darling', a simple table washing machine made by Press Caps Ltd in 1930.

MANGLES AND WRINGERS

Mangling is the process of smoothing and polishing dry or damp-dry laundry by passing it around a roller. The simplest mangling aid was the *board and roller*. Clothes were wrapped around a wooden roller which was placed on a flat surface and rolled backwards and forwards with a heavy board. This process allowed the laundress to put her weight as well as her strength into the polishing operation. Many boards were intricately carved and painted and used as gifts.

The same principle was used in a much heavier piece of equipment, the *box mangle*, developed in the eighteenth century. Working on a large table, the laundress carefully folded smaller items of washing into larger items such as tablecloths and sheets. They were then rolled around a thick wooden roller, and two of these were placed across the bed of the mangle. Over them ran a stout wooden box weighted with stones. A gearing system, or more simply a rope or leather strap, allowed the laundress to crank the box to and fro; the weight of the box was sufficient to smooth and polish damp or dry clothes. While two rollers were being mangled, a third was being loaded on the table.

A box mangle made by Baker of London in 1810 and supplied to the estate laundry at Shugborough, Staffordshire, remained in use until the 1920s. It was operated every Wednesday by a handyman, who was employed once a week to help the two or three laundrymaids.

Another important manufacturer of box mangles was Thomas Bradford of Salford, whose 'Premier' mangle was widely used.

Box mangles were surprisingly effective. The larger items needed no ironing afterwards; the smaller items needed only a little hand finishing. This was an important advantage for large estate

13

laundries or those attached to schools, convents and communal street or village wash-houses. They remained in use well into the twentieth century and still featured in the Army and Navy Stores catalogue of 1907.

Small upright machines for both dry mangling and wet wringing were first patented in the eighteenth century but not manufactured until the mid nineteenth. They were then so successful that they rapidly became almost universal laundry aids. They were cheaper and more popular than washing machines.

Rollers were usually made of pine, beech, maple or some other white wood. Most machines had two rollers but some were three-rollered, in which case the clothes were laid within a holland sheet which was attached to the middle roller; sheets and clothes were carefully wound round the roller while the top and bottom rollers gave constant pressure. By the 1900s many smaller wringers were made

of rubber and did less damage to buttons. Later a 'sandwich' roller with a rubber strip inserted in the middle of the top roller combined the advantages of both wood and rubber.

The regulation of pressure on the rollers was provided by one of three methods, all adapted from cheese presses: a horizontal lever with adjustable weights; elliptical springs and screw; or side-mounted spiral springs and screw.

Early models were simple all-wooden mangles. Later, gearing systems became more sophisticated and the cast-iron frames were elaborately decorated. Ewbank produced a series with names such as the 'Victory', 'Progress', 'Triumph', 'Treasure' and 'Prize'. A useful innovation appeared in the 1900s: hinges on the roller frame enabled it to be folded downwards and the whole wringer converted into a table. Small table-mounted wringers were also available.

The dry laundry at Wormwood Scrubs Prison in the 1890s, showing a box mangle. The laundry dealt with the clothing for 1400 prisoners.

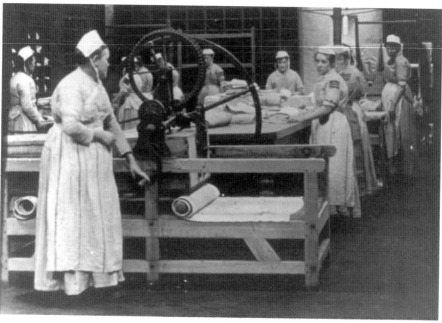

Two views of the ironing and mangling class at the Royal School, Wolverhampton, around 1890. The flat irons were heated on a stove, the top of which can be seen in the lower picture.

Siddons' Fine Cast

With Lignum Vitae Wood Handles

No. 520

Size	No.	Price per doz.
5 Inches	... I ...	54 –
5½ ,,	... 3 ...	66 –
6 ,,	... 5 ...	78 –
6½ ,,	... 7 ...	90 –

No. 520

Heaters

For Fine Cast Box Irons

No.	I	3	5	7	
	5	5½	6	6½	inches
Price		per cwt.			

The most commonly found design of box iron incorporates a vertically sliding door in the heel and was the subject of the first British patent for a box iron, issued to Isaac Wilkinson of Denbighshire in 1738, and later produced in quantity. With slight variations in handle design, Wilkinson's pattern continued to be made throughout the nineteenth and twentieth centuries; accurate dating of individual specimens is extremely difficult. This extract from the catalogue of J. & J. Siddons was advertising the same pattern in 1937.

IRONING

The practice of using heated irons for pressing clothes seems to have originated in China twelve centuries ago. Chinese irons were shaped like small saucepans and finely cast in bronze; used only on silk, they were not common objects.

In Europe in early times creases were removed from clothes by cold smoothing with either a board and roller or calenders made of stone or glass. In Europe the idea of using heat did not develop until the sixteenth century. From this time two types of iron have emerged: the box iron and the sad iron.

Although *box irons* are seemingly the more sophisticated of the two types, the earliest examples of irons are of this kind. Shapes varied but the basic principle remained the same: an iron slug was shaped to fit inside the hollow body of the iron, which had either a hinged or sliding heel or top; only the slug was heated in the fire. They were expensive and early ones were made of elaborately pierced brass or hand-wrought iron; eighteenth-century English irons were usually of iron or steel, sometimes with a layered brass sole. The most commonly found box iron incorporates a vertically sliding door on the heel.

Although many box irons were craftsman-made, a number of large-scale iron manufacturers started production in the eighteenth century, including the firms of Izon and Kenrick.

The laundress needed to buy only one box iron with two iron slugs, which were heated alternately on the open fire. Other advantages of the box iron were its greater cleanliness and fewer fumes.

A variation of the box iron, *charcoal irons* had large deep bodies heated by burning some fuel such as charcoal within the body of the iron. Continental and early English versions allowed the fumes to escape through the holes in the sides of the body, which were often arranged in decorative patterns. In the nineteenth century the shape became standardised and a funnel was introduced which carried the fumes away from the laundered clothes, presumably into the lungs of the laundress! The charcoal, which might be simply embers from the open fire, was

16

Above: *Continental, probably German, chimneyless charcoal iron of a type made in the nineteenth and twentieth centuries, mainly for the Far Eastern market.*

Below: *The 'Dalli' charcoal iron was patented in Germany, but manufactured widely elsewhere under licence throughout the early part of the twentieth century. It used a special compacted charcoal. It was still available in the 1930s and later, costing around 13s plus 3s for a box of fuel.*

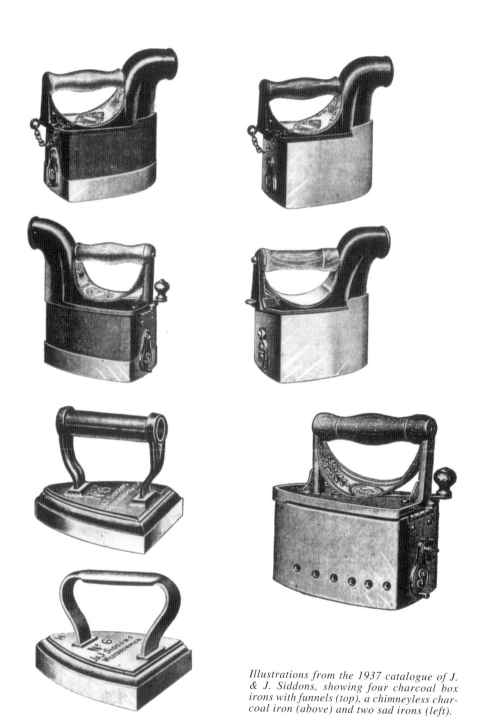

Illustrations from the 1937 catalogue of J. & J. Siddons, showing four charcoal box irons with funnels (top), a chimneyless charcoal iron (above) and two sad irons (left).

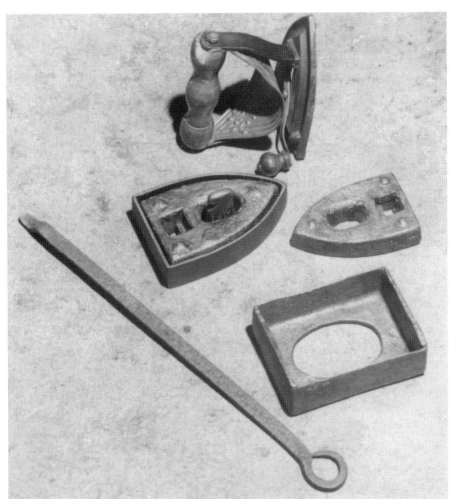

A box iron set made by J. & J. Siddons of West Bromwich. In this case the handle and lid of the iron can be detached from the body; a rectangular container held one of the two heaters near the fire and the set was completed by a poker-like lifter. This design was advertised by Siddons in the 1930s with a choice of bronzed or nickel-plated lids.

kept glowing by opening an aperture in the heel and either blowing on the coals or swinging the iron backwards and forwards.

By the end of the nineteenth century sophisticated charcoal irons were available, with improved ventilation. These charcoal irons were surprisingly light and easy to regulate.

The favourite iron of the nineteenth century, the *sad* (i.e. solid) or *flat iron,* was used in pairs or in threes, one iron being in use while another was heated. They were made of solid cast iron faced with tempered steel. Methods of heating varied according to the situation. Most ordinary householders propped them face up on a trivet in front of the open fire of the domestic range; sometimes the trivet was replaced by a metal hanger suspended from the fire bars. In larger households the tops of the closed

Above: *A gas iron and stand made by the Davis Gas Stove Company in 1930. An unusual feature is the copper hand shield. Lent to the Science Museum, London, by J. F. Parker, Esq.*

Below: *Clark's 'Fairy Prince' enamelled gas iron and stand.*

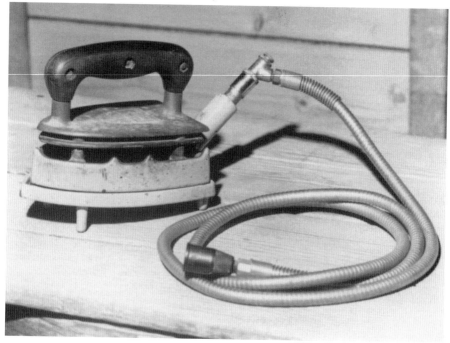

No. 645 "ADONIS" DOMESTIC IRON.

Internally heated. Very highly finished with polished body, Berlin blacked top, hardwood handle, etc. All corners and edges are rounded to save linen. For use with rubber or flexible tubing.

Approximate Weight, ...	7 lbs.
Price, ...	12/9 each.
Nickel-plated Finish, ...	1/9 extra.
If with Nickel-plated Gas and Air Adjuster, ...	1/6 "

No. 647 DOMESTIC IRON.

Highly Polished. Cool Grip. Internally Heated.

Consisting of two Irons and Stand. All corners and edges are rounded to save linen. Can be attached to any gas bracket by using interchangeable tap and flexible tubing. One Iron can be heated while the other is in use. Consumption 4 feet of gas per hour.

Packed in Box ready for delivery.

Price, Two Irons with wood handles
and Stand, complete, ... **£1 5 0** per set.

Extra for Irons Nickel-plated, ... **3/6** per pair.

Mounted on Stands in series of any number.

Extract from a trade catalogue by Carron of Falkirk, 1935, showing the two types of gas iron supplied during the 1920s and 1930s. The gas-stand type was a simple development of the gas-iron heaters that were used fairly commonly for heating flat or goose irons, but it avoided rusting of the sole since the gas jet was directed into the inside of the iron. Gas irons of this type were used in pairs in the same way as flat irons.

ranges were used, although care had to be taken to keep the top of the range free from black lead. Most specialised estate or school laundries invested in a coal-fired laundry stove; in the twentieth century flat irons were also heated over paraffin and gas stoves.

Flat irons were made throughout the nineteenth century by hundreds of foundries throughout the British Isles, in series numbered from one to twelve or fourteen. There was no standardisation in the series, so that similarly numbered irons by different makers can be of different weights. Most English flat irons have handles made of a hollow tube of iron, while box irons often have shaped wooden hand grips.

A type of iron intermediate between the box and the sad iron was patented in the United States of America by Mrs Mary Potts in 1871. This was a double-pointed iron with a detachable handle and sold in sets of three bodies, one handle and a stand. In England A. Kenrick of West Bromwich made irons of this type under licence from 1880 onwards.

More conventionally shaped detachable-handle irons are also occasionally found and were manufactured throughout the nineteenth century.

During the second half of the nineteenth century and the early twentieth, a surprising variety of fuels was used before electricity finally solved the problem of heat supply.

The first paraffin irons appeared in the 1890s. Petroleum was used as an iron fuel in both Europe and the United States, and by 1900 A. Kenrick & Sons were supplying the 'Bonus' petrol iron, similar in shape to the Potts iron but with a conventional handle. Naphtha was used in an American patent of 1868. Vegetable oil (colza or rape seed) was used in French irons; the oil was poured into a lip around the sole and lit. Methylated spirit was a more widely used fuel, especially in travelling irons.

Coal gas was introduced as a means of heating irons in the early twentieth century. Two basic types were produced. One was heated by a burner fixed inside the body of the iron and fed through a

An electric iron patented in 1891 and made by Cromptons. Although typically nineteenth-century in appearance, it was heated by an internal resistance element similar to that used in modern irons.

flexible tube. The great disadvantage was the production of fumes, which were inhaled by the laundress. An unusual feature of gas irons of this type is the single-post handle that was incorporated into the design of many of them.

The other type of gas iron was hollow-bodied with a wide opening in the heel, which was slotted over a burner incorporated into a stand.

The first electric irons were introduced in the 1880s in the United States, but these early electric irons were cumbersome and rare. One of the earliest and most dangerous models was made in France and was heated by an electric arc between two adjustable carbon electrodes. By 1907 the Army and Navy Stores in London were advertising electric laundry irons weighing 6 pounds (2.7 kg) for 22 shillings. Compare this cost with the price of older types of iron in the same catalogue – box irons from 10d to 3s 7d – and one understands why the use of traditional irons persisted so long. It was not until the 1920s and 1930s that electric irons became more popular.

SPECIALISED IRONS

There were a number of specialised irons, which are occasionally found today. Many are illustrated in the extract from Kenrick's catalogue of 1926 (page 25).

Travelling irons needed to be light, small and easy to heat. In the 1890s Josef Feldmeyer of Wurzburg patented a spirit-heated iron of a design manufactured in many European countries under different names. Other travelling ironing kits were fuelled by alcohol, petrol, electricity or patent solid fuel.

A very heavy, elongated flat iron, known as a *tailor's goose*, was used for pressing tailored seams. It has a distinctive solid handle with a twist grip. Weights ranged from 7 to 28 pounds (3.2 to 12.7 kg).

Convex-bottomed *glossing irons* were heated to a very high temperature and used for polishing starched linen and cotton. Glazing could also be accomplished by the more primitive method of rubbing with a highly polished 'slick' stone such as an agate or with a shaped piece of glass, or even with the bottom of

Above: *The 'Brilliant' spirit-heated travelling iron and stand, early twentieth century. This was manufactured to Feldmeyer's design and sold for 6s in 1907.*

Below: *The 'Boudoir Supreme' travelling iron and curling-tongs heater with its patent Meta fuel, 1914. Lent to the Science Museum, London, by Miss C. Gallacher.*

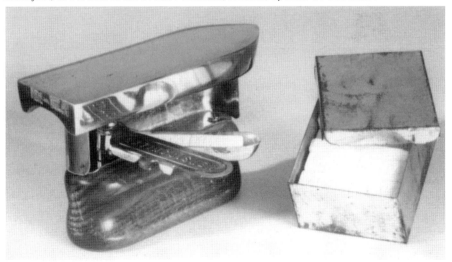

GEC travelling iron, 1922. When upturned on its stand, the sole of the iron could be used to heat water in the container for either shaving or an early morning cup of tea.

a glass bottle.

Double-pointed flat irons, when turned around, described a circle. They were used by milliners for ironing the crowns and brims of bonnets and caps. A deep-bodied blunt iron called a *shell* was also used by hatters.

Lace irons were either single or double pointed box or sad irons, but smaller than irons for ordinary laundry work.

Also called *flounce* or *ox-tongue irons*, descriptive of their long curved body, *sleeve irons* were used for reaching high up into pleats or sleeves. Another type of sleeve iron was mounted on a long handle.

Heavy rectangular solid irons were used for pressing the baize of billiards tables. They were usually supplied with a light metal shoe to protect the sole.

STANDING IRONS

There survive a number of different types of iron mounted on stands or long handles and used in an upright position for specialised and highly skilled finishing. They include egg irons, cap crown irons and Italian irons.

The *egg iron* was a solid egg-shaped piece of iron mounted on either a wooden or metal stand and used for finishing the tops of sleeves or gathered waistbands.

The *cap crown iron* had a circular or half-circular top of iron or brass, sometimes with a protective cap of brass that fitted over the iron after heating. Eighteenth-century versions have extremely elegant stands, sometimes decoratively turned from two different coloured woods.

The *Italian iron* was a hollow cylinder of brass or iron heated by two alternating pokers, which were wedged into the open fire. Like other upright standing irons, eighteenth-century examples were often finely made of brass and mahogany or steel, sometimes with more than one cylinder to each stand; these were found only in well-to-do households. In the nineteenth and early twentieth centuries plainer cast-iron versions were mass-produced.

Known in Elizabethan times, Italian irons were used for finishing and crimping fine lace edges and frills. They remained in use throughout the nineteenth and early twentieth centuries but were used then mainly for finishing bows and

24

DOMESTIC IRONMONGERY.

No. 700, Sad Iron. Convex Sad Iron Flounce Iron. One Set of " A " Polished Mrs. Pott's Patent Cold-handle Sad Irons.

Egg Iron. Sleeve Iron. Lace Iron, complete with Stand.

Broad Tailor's Iron, with Bevelled Point. Tailor's Iron, with Square Point. Billiard Table Iron.

Hatter's Rim Iron, No. 1. Hatter's Iron, No. 2. Hatter's Iron, No. 3.

Hatter's Rim Iron, No. 4. Hatter's Rim Iron, No. 5. Hatter's Rim Iron, No. 6.

Extract from a trade catalogue of A. Kenrick & Sons Ltd, West Bromwich, 1926. Ordinary sad irons and the tailor's goose irons were supplied to retailers by the hundredweight. Less usual irons such as the convex polishers and the lace irons were priced in dozens and ranged from 26s to 70s (wholesale) per dozen best-quality nickel-plated. Speciality items such as flounce irons (6s 3d to 9s 4d), sleeve irons (2s 6d to 3s 2d), hatter's irons (3s 2d to 8s), rim irons (3s 2d to 3s 6d) and billiards-table irons (5s to 10s) were priced individually. The Potts cold-handle sad irons were sold as a set priced between 10s and 12s (wholesale).

25

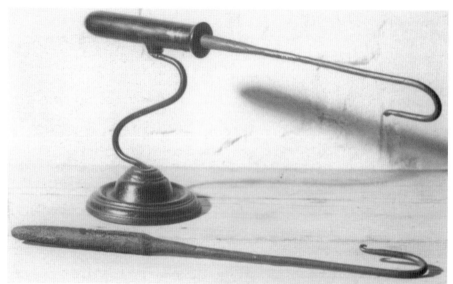

Above: *A mass-produced nineteenth-century Italian finishing stand with two heating irons to be used alternately. Stands of this type had plain irons with looped handles.*

Below left: *The more elegant and elaborate hand-made stands of previous centuries had irons with decorative loops, twirls or turned wooden knobs, with complicated designs cut or twisted into the length of the iron. The stands themselves had wooden or trivet bases. Occasionally the base can be used to clamp the stand to the edge of a table.*

Below right: *A goffering stack. The starched wet linen or cotton was threaded through the spills and dried by the fire. The frame could then be carefully dismantled.*

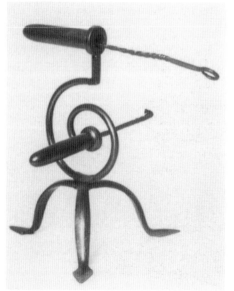

Crimping boards and rollers. The linen was laid on the board and the fluted roller was then rolled over it a number of times and the ridges produced in the linen were sewn together, forming the very even and finely gathered finish so often seen in early children's wear.

awkward ribbons and for ironing velvets or other pile fabrics used in elaborate Victorian and Edwardian dresses.

Caps, collars, cuffs and ruffles needed crimping and goffering. Both processes gave a corrugated finish but crimping produced a tighter, finer effect. It was used in making the very fine gathers on waistbands and shirt sleeves. The earliest aids were the *crimping frame* or *goffering stack*, and the *board and roller*. In the mid nineteenth century other fluting aids were invented, such as *goffering tongs*, scissor-like tongs with long, plain, cylindrical blades. They were heated over a naked flame and tested on paper. *Crimping tongs* were similar but had shorter, wider, fluted blades. Obtainable at a high price from 1860 onwards were miniature fluted wringers, finely made in brass. Some flat irons carried a fluting surface mounted on the side.

A late nineteenth-century crimping machine.

27

The 'Tower' storeyed flat-iron stove, manufactured by Thomas Bradford & Company, Salford, and supplied to the Royal School, Wolverhampton. The stove will heat fifty irons, including four polishers, and is now in the collection of the Staffordshire Arts and Museum Service.

Three iron stands from the collections of the Staffordshire Arts and Museum Service. Of particular interest is the stand made from a rough piece of wood and a patten ring. Pattens were devices strapped over shoes to raise the wearer out of the wet and were used by laundresses.

ACCESSORIES AND FITTINGS

Three-legged *iron stands* were made in a wide variety of shapes and decoration and vary from the finely made brass trivets of the eighteenth century to simple but effective home-made stands using domestic cast-offs such as horseshoes.

Pairs of flat irons could be heated, heels down, on a wire *iron hanger*, which hooked over the fire bars.

Long-handled *charcoal ladles* made of wire were used to hold charcoal as it heated in the fire.

Iron shoes were metal plates with spring wire clips to fit over the sole of the iron. The flat iron was placed in the shoe immediately after heating to keep the sole clean. This device was invented in the eighteenth century but mass-produced versions became commonly available at the end of the nineteenth century.

A piece of cloth was usually needed to protect the hand when using hot flat irons. These *iron holders* could be made out of old pieces of blanket tacked together, with brown paper in between to help insulation.

Wooden-screw *linen presses* were used for the storage of newly mangled laundry. They were made of pine, oak, fruitwood or mahogany, either to stand on top of a chest of drawers or, more rarely, incorporating drawers underneath the press. Attractive polished presses were marketed by Thomas Bradford & Company of Salford, one of the most important manufacturers and suppliers of laundry and dairy equipment in the late nineteenth century.

Coal-fired *stoves* for heating flat irons were produced throughout the second half of the nineteenth century. They vary in type, some being flat-topped, as glimpsed in the lower photograph on page 15, others being storeyed, as marketed by Thomas Bradford of Salford under the names of the 'Pagoda' and the 'Tower'. Size varied from stoves for eight or twelve irons to ones for sixty-two. Tailor's stoves

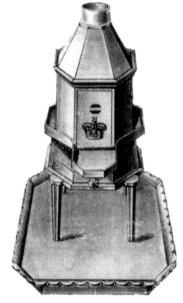

Illustrations from the trade catalogue of Taylors of Driffield, 1897, showing, from the left, an ironing stove, a laundry stove, and a 'Patent Economical' laundry stove.

were also made to accommodate the long-bodied goose iron, which later could also be heated on single- or double-burner gas heaters.

Large laundries incorporated built-in *laundry tables* solidly made in pine. Tapered skirt boards supported on trestles were used in pressing long full skirts, and smaller tapered boards for sleeves. Rectangular boards, approximately 18 by 12 inches (450 by 300 mm), were used

for ironing shirt fronts. All these boards needed to be covered first by an ironing blanket, usually a piece of old woollen material or, in large laundries, a specially felted, thickly woven blanket, and then by a calico ironing sheet. The reverse side of the shirt board could be kept uncovered and used as a base when polishing the starched front with a polishing iron. Shaped boards were also used for polishing caps and collars.

FURTHER READING

Army and Navy Stores. *Yesterday's Shopping*. David & Charles, reprint of 1907 catalogue.
Davidson, Caroline. *A Woman's Work Is Never Done*. Chatto & Windus, 1982.
Filbee, Marjorie. *A Woman's Place*. Ebury Press, 1980.
Hardyment, Christina (editor). *The Housekeeping Book of Susannah Whatman, 1776–1800*. Century, 1987.
Hardyment, Christina. *From Mangle to Microwave*. Polity Press, 1988.
Hardyment, Christina. *Behind the Scenes: Domestic Arrangements in Historic Houses*. National Trust, 1997. (Previously published as *Home Comforts: A History of Domestic Arrangements*, Viking in association with the National Trust, 1992.)
Malcolmson, Patricia E. *English Laundresses: A Social History, 1850–1930*. University of Illinois Press, 1986.
Sambrook, Pamela A. *The Country House Servant*. Sutton Publishing in association with the National Trust, 1999.
Sewell, Brian. *Smoothing Irons*. Midas Books, 1977.
Seymour, John. *National Trust Book of Forgotten Household Crafts*. Dorling Kindersley, 1987.
Thomson, Ruth. *Washday*. A. & C. Black, 1990.
Walkley, C., and Foster, V. *Crinolines and Crimping Irons*. Peter Owen, 1978.
Weaver, Rebecca, and Dale, Rodney. *Machines in the Home*. The British Library, 1992.

The 'Big Ben' tailor's iron stove supplied to a small jobbing tailor in Burton upon Trent in 1912. Now in the collection of the Staffordshire Arts and Museum Service.

PLACES TO VISIT

Museum displays may be altered and readers are advised to telephone before visiting to check that relevant items are on show, as well as to find out the opening times.

Anne of Cleves House, 52 Southover High Street, Lewes, East Sussex BN7 1JA. Telephone: 01273 474610. Website: www.sussexpast.co.uk

Audley End House, Audley End, Saffron Walden, Essex CB11 4JG. Telephone: 01799 522399. Website: www.english-heritage.org.uk

Beningbrough Hall, Beningbrough, York YO30 1DD. Telephone: 01904 470666. Website: www.nationaltrust.org.uk

Berrington Hall, Leominster, Herefordshire HR6 0DW. Telephone: 01568 615721. Website: www.nationaltrust.org.uk/berrington

Bygones at Holkham, Holkham Park, Wells-next-the-Sea, Norfolk NR23 1AB. Telephone: 01328 710227. Website: www.holkham.co.uk

Cambridge and County Folk Museum, 2/3 Castle Street, Cambridge CB3 0AQ. Telephone: 01223 355159. Website: www.folkmuseum.org.uk (Museum closed for refurbishment until early 2005.)

Castle Ward, Strangford, Downpatrick, County Down, Northern Ireland BT30 7LS. Telephone: 028 4488 1204. Website: www.nationaltrust.org.uk

Church Farm Museum, Church Road South, Skegness, Lincolnshire PE25 2HF. Telephone: 01754 766658. Website: www.lincolnshire.gov.uk/churchfarmmuseum

Dunham Massey, Altrincham, Cheshire WA14 4SJ. Telephone: 0161 941 1025. Website: www.nationaltrust.org.uk

Erddig, Wrexham LL13 0YT. Telephone: 01978 355314. Website: www.nationaltrust.org.uk

How We Lived Then, 20 Cornfield Terrace, Eastbourne, East Sussex BN21 4NS. Telephone: 01323 737143. Website: www.how-we-lived-then.co.uk

Killerton, Broadclyst, Exeter, Devon EX5 3LE. Telephone: 01392 881345. Website: www.nationaltrust.org.uk

Kingston Lacy, Wimborne Minster, Dorset BH21 4EA. Telephone: 01202 883402. Website: www.nationaltrust.org.uk

Llanerchaeron, Ciliau Aeron, near Aberaeron, Ceredigion SA48 8DG. Telephone: 01545 570200. Website: www.nationaltrust.org.uk

Milton Keynes Museum, McConnell Drive, Wolverton, Milton Keynes MK12 5EL. Telephone: 01908 316222. Website: www.mkmuseum.org.uk

Museum of Lincolnshire Life, Burton Road, Lincoln LN1 3LY. Telephone: 01522 528448. Website: www.lincolnshire.gov.uk/museumoflincolnshirelife

Museum of Welsh Life, St Fagans, Cardiff CF5 6XB. Telephone: 029 2057 3500. Website: www.nmgw.ac.uk

Ormesby Hall, Church Lane, Ormesby, Middlesbrough TS7 9AS. Telephone: 01642 324188. Website: www.nationaltrust.org.uk

Pickford's House, 41 Friar Gate, Derby DE1 1DA. Telephone: 01332 255363. Website: www.derby.gov.uk

The Science Museum, Exhibition Road, South Kensington, London SW7 2DD. Telephone: 0870 870 4868. Website: www.sciencemuseum.org.uk

Staffordshire Arts and Museum Service, County Museum, Shugborough, Stafford ST17 0XB. Telephone: 01889 881388. Website: www.staffordshire.gov.uk

Waterperry Country Life Museum, Waterperry Gardens, near Wheatley, Oxfordshire OX33 1JZ. Telephone: 01844 339226. Website: www.waterperrygardens.co.uk

Wayside Museum, Zennor, St Ives, Cornwall TR26 3DA. Telephone: 01736 796945.

Worcestershire County Museum, Hartlebury Castle, Hartlebury, Kidderminster, Worcestershire DY11 7XZ. Telephone: 01299 250416. Website: www.worcestershire.gov.uk

York Castle Museum, The Eye of York, York YO1 9RY. Telephone: 01904 687687. Website: www.yorkcastlemuseum.org.uk